ASSESSMENT HANDBOOK

Everyday Mathematics®
The University of Chicago School Mathematics Project

The University of Chicago School Mathematics Project (UCSMP)

Max Bell, Director, *Everyday Mathematics* First Edition; James McBride, Director, *Everyday Mathematics* Second Edition; Andy Isaacs, Director, *Everyday Mathematics* Third and Fourth Editions; Amy Dillard, Associate Director, *Everyday Mathematics* Third Edition; Rachel Malpass McCall, Associate Director, *Everyday Mathematics* Fourth Edition; Mary Ellen Dairyko, Associate Director, *Everyday Mathematics* Fourth Edition

Authors
Jean Bell, Max Bell, David W. Beer*, Dorothy Freedman, Nancy Guile Goodsell†, Nancy Hanvey, Deborah Arron Leslie, Kate Morrison

* Third Edition only
† First Edition only

Fourth Edition Kindergarten Team Leader
Deborah Arron Leslie

Writers
Kathryn Flores, Gina Garza-Kling, Allison M. Greer, Amanda Louise Ruch, Elizabet Spaepen

Differentiation Team
Ava Belisle-Chatterjee, Leader; Jean Marie Capper

Digital Development Team
Carla Agard-Strickland, Leader; John Benson, Gregory Berns-Leone, Juan Camilo Acevedo

Virtual Learning Community
Meg Schleppenbach Bates, Cheryl G. Moran, Margaret Sharkey

Technical Art
Diana Barrie, Senior Artist; Cherry Inthalangsy

UCSMP Editorial
Molly Potnick, Loren Santow

Field Test Teachers
Rebecca Criollo, Mandy Djikas, Megan DeBetta, Pamela Downing, Tamela Fralin, Laney Frazier, Eric Lester, Priscilla Lindsey, Kari E. Moulton, Nichole Parmley, Mary Rogers

Digital Field Test Teachers
Colleen Girard, Michelle Kutanovski, Gina Cipriani, Retonyar Ringold, Catherine Rollings, Julia Schacht, Christine Molina-Rebecca, Monica Diaz de Leon, Tiffany Barnes, Andrea Bonanno-Lersch, Debra Fields, Kellie Johnson, Elyse D'Andrea, Katie Fielden, Jamie Henry, Jill Parisi, Lauren Wolkhamer, Kenecia Moore, Julie Spaite, Sue White, Damaris Miles, Kelly Fitzgerald

Contributors
Ann E. Audrain, John Benson, Patrick Carroll, Andy Carter, Jeanne Mills DiDomenico, James Flanders, Margaret Krulee, Lila K.S. Goldstein, Barbara Smart, Penny Williams

Center for Elementary Mathematics and Science Education Administration
Martin Gartzman, Executive Director; Meri B. Fohran, Jose J. Fragoso, Jr., Regina Littleton, Laurie K. Thrasher

External Reviewers

The *Everyday Mathematics* authors gratefully acknowledge the work of the many scholars and teachers who reviewed plans for this edition. All decisions regarding the content and pedagogy of *Everyday Mathematics* were made by the authors and do not necessarily reflect the views of those listed below.

Elizabeth Babcock, California Academy of Sciences; Arthur J. Baroody, University of Illinois at Urbana-Champaign and University of Denver; Dawn Berk, University of Delaware; Diane J. Briars, Pittsburgh, Pennsylvania; Kathryn B. Chval, University of Missouri–Columbia; Kathleen Cramer, University of Minnesota; Ethan Danahy, Tufts University; Tom de Boor, Grunwald Associates; Louis V. DiBello, University of Illinois at Chicago; Corey Drake, Michigan State University; David Foster, Silicon Valley Mathematics Initiative; Funda Gönülateş, Michigan State University; M. Kathleen Heid, Pennsylvania State University; Natalie Jakucyn, Glenbrook South High School, Glenview, IL; Richard G. Kron, University of Chicago; Richard Lehrer, Vanderbilt University; Susan C. Levine, University of Chicago; Lorraine M. Males, University of Nebraska-Lincoln; Dr. George Mehler, Temple University and Central Bucks School District, Pennsylvania; Kenny Huy Nguyen, North Carolina State University; Mark Oreglia, University of Chicago; Sandra Overcash, Virginia Beach City Public Schools, Virginia; Raedy M. Ping, University of Chicago; Kevin L. Polk, Aveniros LLC; Sarah R. Powell, University of Texas at Austin; Janine T. Remillard, University of Pennsylvania; John P. Smith III, Michigan State University; Mary Kay Stein, University of Pittsburgh; Dale Truding, Arlington Heights District 25; Judith S. Zawojewski, Illinois Institute of Technology

Note
Too many people have contributed to earlier editions of *Everyday Mathematics* to be listed here. Title and copyright pages for earlier editions can be found at http://everydaymath.uchicago.edu/about/ucsmp-cemse/.

www.everydaymath.com

Copyright © McGraw-Hill Education

All rights reserved. The contents, or parts thereof, may be reproduced in print form for non-profit educational use with Everyday Mathematics, provided such reproductions bear copyright notice, but may not be reproduced in any form for any other purpose without the prior written consent of McGraw-Hill Education, including, but not limited to, network storage or transmission, or broadcast for distance learning.

Send all inquiries to:
McGraw-Hill Education
8787 Orion Place
Columbus, OH 43240

ISBN: 978-0-07-902114-4
MHID: 0-07-902114-X

Printed in the United States of America.

6 BRP 23

Contents

Assessment in *Everyday Mathematics* 1

Interim Assessments

Overview 5

Beginning-of-Year Assessment Masters . 11

Mid-Year Assessment Masters 14

End-of-Year Assessment Masters 23

Record-Keeping Masters

Blank Section Checklists 35

Blank Mathematical Process and
   Practice Checklists 37

Mathematical Process and Practice in
   Open Response Problems Checklist . 40

General Masters

My Exit Slip 41

Good Work! 42

My Work 43

About Math Time 44

Appendix

GMP Rubrics A1

Assessment in Everyday Mathematics®

Assessment in *Everyday Mathematics*:

- addresses the full range of content and processes/practices in the Standards for Mathematics.
- consists of tasks that are worthwhile learning experiences.
- is manageable for teachers.
- informs instruction by providing actionable information about children's progress.
- provides information for grading.
- clarifies the *Everyday Mathematics* spiral and helps teachers decide when to intervene and when "watchful waiting" is appropriate.
- serves basic Tier 1 and Tier 2 Response to Intervention (RtI) functions.
- provides information that will complement data from standards-based assessments, including those from PARCC and SBAC.

(Go Online) for more information in the Assessment section of the *Implementation Guide*.

Assessment of Content and Process/Practice Standards

Everyday Mathematics integrates instruction and assessment of mathematical processes and practices with instruction and assessment of grade-level content. The mathematical processes and practices are not to be separated from the content; they are mathematical habits of mind children should develop as they learn mathematical content.

However, the content and process/practice standards differ in important respects. The content standards describe specific goals that are organized by mathematical strand and differ from grade to grade. The process and practice standards describe general, cross-grade goals that are related to processes and practices such as problem solving, reasoning, and modeling. Many tasks in *Everyday Mathematics* provide opportunities to assess both content and process/practice standards. However, due to the differing nature of these standards, *Everyday Mathematics* assesses and tracks progress on them in different ways.

Assessing the Content Standards

Each grade's content standards are unpacked into 45 to 80 *Everyday Mathematics* Goals for Mathematical Content (GMC). The standards and the corresponding GMCs are listed in the back of the *Teacher's Lesson Guide*. Instructional activities and assessment items are linked to one or more of the GMCs, which provide more targeted information for assessment and differentiation.

For each task that assesses a content standard, *Everyday Mathematics* provides guidance on what constitutes "meeting expectations" for that standard at that point in the year. Individual Profiles of Progress, Class Checklists, and the assessment and reporting tools help teachers monitor children's progress using this framework.

Assessing the Process and Practice Standards

Since the Standards for Mathematical Process and Practice are broadly written for Grades K to 12, *Everyday Mathematics* includes Goals for Mathematical Process and Practice (GMP) that unpack the standards for elementary school teachers and children. These *Everyday Mathematics* GMPs are useful for assessing the mathematical process and practice standards because they highlight specific aspects of each process and practice.

Tracking progress on the mathematical process and practice standards requires a more qualitative approach. Assessment opportunities include open response problems and observations of children working on activities and problems, using manipulatives, and discussing solutions in the course of their daily work. Tools for assessing the processes and practices include checklists and task-specific rubrics for open response problems.

Go Online for more information about the GMCs and GMPs in the *Everyday Mathematics* and the Standards section in the *Implementation Guide*.

Assessment Opportunities

Everyday Mathematics offers a variety of opportunities for ongoing and periodic assessment of content and process/practice standards. Every Section Organizer includes information about assessment opportunities in the section.

Assessment Check-Ins

Assessment Check-Ins are lesson-embedded opportunities to assess the focus content and processes/practices of the lesson. They appear in regular lessons and Open Response and Reengagement lessons. Each Assessment Check-In provides

✓ **Assessment Check-In** K.CC.1

As they play the game, note children's counting accuracy and automaticity. Do they say the next number correctly? How much think time do they require? Expect most children to be able to orally count accurately and efficiently from 1 to at least 10. Many children will be able to count higher. Use your observations to plan for verbal counting practice in upcoming lessons and for playing *Count and Sit* variations in small groups or as a whole group. GMP6.4

Assessment and Reporting Go Online to record children's progress and to see trajectories toward mastery for these standards.

information on expectations for particular standards at that point in the curriculum. The results can be used to inform instruction and, often, for grading.

Interim Assessments

These assessments are administered at the beginning, middle, and end of the school year. For Kindergarten, most of the tasks are best administered in a one-on-one or small-group format, although some can be done as whole-class activities. See pages 5–7 for more information.

- **Beginning-of-Year Assessment** The Beginning-of-Year Assessment provides information about children's knowledge and skills upon entering Kindergarten. The tasks primarily relate to the content in the first two or three sections of the Kindergarten *Teacher's Lesson Guide*. The Beginning-of-Year Assessment can be useful for RtI screening.

- **Mid-Year Assessment** The Mid-Year Assessment provides an opportunity for teachers to assess children's performance on the content from the first five sections of the Kindergarten *Teacher's Lesson Guide*. It provides a fairly comprehensive snapshot of whether children are meeting expectations for the standards covered to date.

- **End-of-Year Assessment** The End-of-Year Assessment offers teachers an opportunity to assess children's performance on all of the Kindergarten Standards. Teachers may choose to use some or all of the items on this assessment, depending on whether they want to check in on all of the standards, or just those for which they lack adequate information from other sources, such as the Mid-Year Assessment and Assessment Check-Ins. All tasks on the Mid-Year and End-of-Year Assessment are appropriate for grading because they match the expectations for the standards they assess up to that point in the year.

Other Assessment Opportunities

Almost any task in *Everyday Mathematics* can provide information that could be useful for assessment. Assessment tools and systems in *Everyday Mathematics* can accommodate data from sources other than those listed above, and teachers should use their judgment about expanding the range of data they gather and use to inform their instruction and assign grades.

Assessment Tools

Everyday Mathematics provides a variety of tools for collecting, storing, analyzing, reporting, and using assessment data.

Rubrics for Evaluating Mathematical Process and Practice Standards

The Open Response and Reengagement lessons are particularly powerful opportunities to assess children's progress on the Standards for Mathematical Process and Practice. Each of these includes a task-specific rubric that can be used to evaluate children's work on the open response problem for a specific *Everyday Mathematics* Goal for Mathematical Process and Practice.

Sample Rubric

Goal for Mathematical Process and Practice GMP7.1 Look for mathematical structures such as categories, patterns, and properties.	Not Meeting Expectations	Partially Meeting Expectations	Meeting Expectations	Exceeding Expectations
	Does not show evidence of identifying or using a pattern in any solutions (see Meeting Expectations).	Identifies one or two instances of a pattern (see Meeting Expectations), but does not generalize to any more solutions.	Shows evidence of or can explain at least one of the following solution patterns: • Turn-around pattern • +1, −1 pattern	Meets expectations, and shows evidence of using or being able to explain another pattern.

Go Online for generic rubrics for the Goals for Mathematical Process and Practice that you can complete and use to evaluate children on the processes and practices highlighted in the Assessment Check-Ins or embedded in the lessons themselves. For information on how to use the rubrics, see the Assessment section in the *Implementation Guide*.

Individual Profiles of Progress and Class Checklists

Individual Profiles of Progress (IPPs) combine data from various sources for individual children. Class Checklists facilitate collecting and recording data for an entire class. Blank masters of these forms are provided in this handbook.

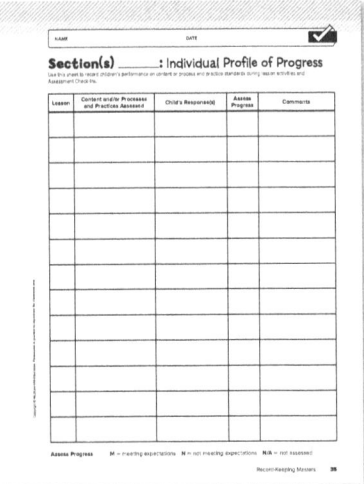
Individual Profile of Progress

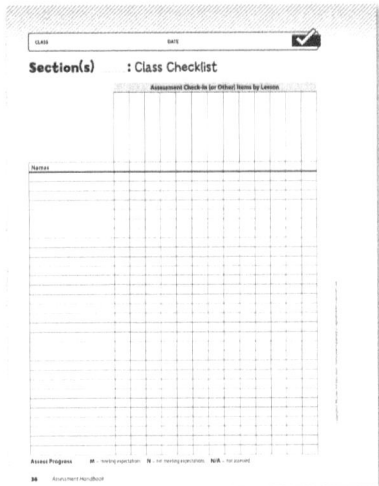
Class Checklist

Go Online for section-specific versions of the IPPs and Class Checklists that you can download and print or complete digitally.

Assessment and Reporting Tools

Digital tools available to teachers through McGraw-Hill's ConnectED platform centralize evaluation, reporting, and targeted differentiation. Teachers are able to evaluate children's work and generate reports on children's progress based on standards covered in lessons and sections. The system can track children's performance and provide teachers information and access to materials at point of use to support differentiation decisions.

Interim Assessments

Overview and Timeline for Interim Assessments

Interim assessment activities are scheduled at regular intervals during the school year to help teachers add to the assessment picture that is generated from ongoing assessment activities, including the Assessment Check-Ins. While interim assessments for older children often focus on paper-and-pencil tasks, interim assessments in *Kindergarten Everyday Mathematics* focus on low-key interactions with the teacher and hands-on tasks, such as working with manipulatives or playing games. Most of these interim assessment tasks are best done with individual children or in small groups, but some can be integrated as part of whole-class activities.

Beginning-of-Year (BOY) Assessment

The tasks on the Beginning-of-Year Assessment relate to topics that you will cover in the first few sections of the *Teacher's Lesson Guide*. Conduct these tasks during the first few weeks of the school year (once children have settled in to classroom routines) to help you:

- know your children's entering levels of skill and understanding, and
- plan instruction accordingly during the first few sections of the *Teacher's Lesson Guide*.

See pages 11–13 for the Beginning-of-Year Assessment.

Mid-Year (MY) and End-of-Year (EOY) Assessments

Conduct the Mid-Year Assessment after you have completed Section 5 in the *Teacher's Lesson Guide,* and conduct the End-of-Year Interim Assessment at the end of the year. These assessments will help you gather data about children's progress toward the content standards and compare children's performance against expectations for the standards at that time of year. The assessment sheets on pages 14–22 (Mid-Year) and pages 23–34 (End-of-Year) include space to record whether children are meeting these expectations, as well as their specific responses, the strategies they use, and areas where they demonstrate difficulty or confusion. These records will help you use the interim assessment information to:

- plan and tailor instruction,
- complete report cards and conduct parent conferences,
- communicate with next-grade teachers, and/or
- report on progress toward school, district, or state expectations.

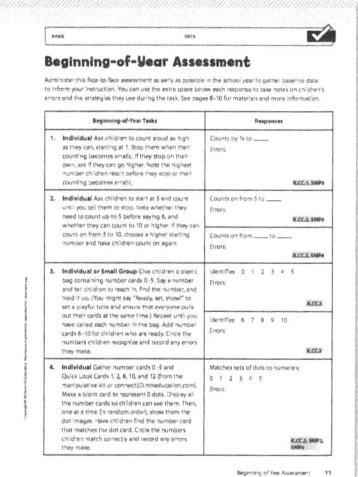

Conduct the Beginning-of-Year Assessment at the beginning of the school year.

You can spread your interim assessment tasks over a period of several weeks according to the details of your particular schedule.

See pages 14–34 for the Mid-Year and End-of-Year Assessments.

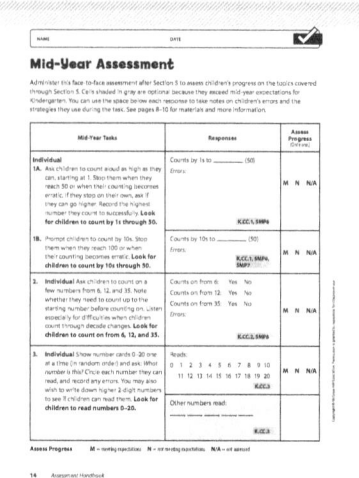

Conduct the Mid-Year Assessment after Section 5 in the *Teacher's Lesson Guide*.

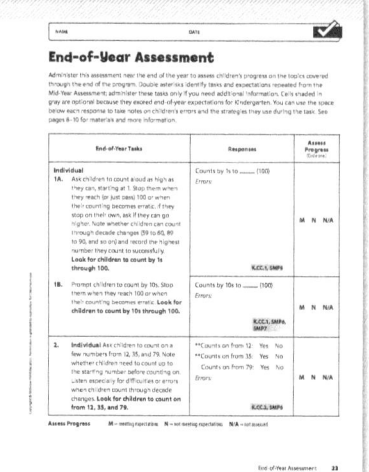

Conduct the End-of-Year Assessment at the end of the school year.

NOTE Many skills or concepts are addressed on more than one of the interim assessments (Mid-Year and End-of-Year, for example). Choose or modify tasks according to your assessment needs. If you have already gathered adequate information about a particular skill or concept from the Assessment Check-Ins or a previous interim assessment, you may not need to conduct tasks on those topics for particular children or for the whole class.

Individualizing Interim Assessments

The interim assessment tasks in this section are intended to help you learn as much as possible about what individual children understand and can do at a particular point in time. To maximize the information you can learn, you may need to modify or adapt the tasks according to the child or group of children you are working with at that moment. For example:

- If a child has difficulty with a task, simplify it slightly or engage the child in conversation about the task to better understand the root of the difficulty.

- If a child performs an activity with ease, add a bit of challenge to see how much farther he or she can go.

- Allow children to take open-ended tasks as far as they are able. Provide encouragement for children to try things, even if they think they are difficult.

- If they seem perplexed by a question or set of instructions, try presenting the information in a different way to see if it makes more sense to them. Children with special needs or learning differences may require specific modifications to help them best express what they know.

In all cases, keep in mind that the interim assessment tasks are designed to be implemented flexibly, rather than administered verbatim.

NOTE Using an approach that goes beyond simply checking correct or incorrect, or meets or does not meet expectations, may require a bit more time and energy on your part. However, it will pay off in the increased information you will gather to help you tailor your instruction and communicate with parents or other teachers. You may find that this approach allows the assessment opportunities to become rich interactions for teaching and learning as well.

Tips for Implementing Interim Assessments

Once you decide which tasks you will use from the interim assessments, consider how you will administer them, keeping in mind that they don't need to be done all at once for a particular child and that they don't need to be done in the order in which they are written. You may wish to cluster the one-on-one tasks into a single sitting during which you meet with the child in the hallway or a quiet corner of the classroom. Other tasks can be administered with small groups during a Center or other work time over the course of several days. Consider the following tips for scheduling and implementing tasks that you plan to do with individuals or small groups:

- Find a place where you and children can hear each other easily and where children can concentrate.
- Arrange for parents or other adults to help in the classroom so you can devote your attention to the assessment activities.
- Set a playful tone, rather than a pressured one. You might explain to children that everyone will have a turn to do math activities and games with you over the course of several days (or longer).
- In advance, collect all of the materials and recording sheets that you will need to do all of the tasks you have planned.

See the Overview table that follows for a summary of the tasks on each interim assessment, as well as the Standards they address, the suggested grouping formats, and the materials needed.

Assessment Overview

Task # BOY	Task # MY	Task # EOY	Content Assessed	Possible Formats/ Groupings*	Materials Needed
1	1	1	oral counting **K.CC.1, SMP6, SMP7**	I	none
2	2	2	counting on **K.CC.2, SMP6**	I	none
3	3	3	reading numerals **K.CC.3**	I, SG	**Beginning-of-Year** Individual Number Cards 0–5 (*Math Masters*, p. TA34); plastic bags; Individual Number Cards 6–10 (*Math Masters*, pp. TA34–TA35) (optional) **Mid-Year, End-of-Year** Individual Number Cards 0–20 (*Math Masters*, p. TA34–TA37)
4			matching sets and numerals **K.CC.3, SMP2, SMP6**	I	Individual Number Cards 0–5 (*Math Masters*, p. TA34); Quick Look Cards 1, 2, 8, 10, 12; blank card representing 0 dots
	4	4	writing numerals **K.CC.3**	I, SG, WC	completed number scrolls or Blank Number Scrolls (*Math Masters*, p. TA52)
5	5		counting principles **K.CC.4a, K.CC.4b, K.CC.4c, K.CC.5, SMP6**	I	7 connecting cubes (or other small objects); cup
	6	5	matching sets and numerals, counting sets **K.CC.3, K.CC.5, SMP2, SMP6**	I	pre-made sets of 5, 12, 20, 9, and 17 counting bears (or other small objects); Individual Number Cards 0–20 (*Math Masters*, pp. TA34–TA37)
6	7	6	counting out sets **K.CC.5, SMP6**	I	bag of 20–30 connecting cubes (or other small objects)
7	8	7	comparing sets **K.CC.6, SMP2, SMP6**	I, SG	**Beginning-of-Year** 7 connecting cubes (5 of one color, 2 of another color) **Mid-Year, End-of-Year** *Top-It* Dot Cards (*Math Masters*, pp. G3–G7)
	9	8	comparing numerals **K.CC.7, SMP2, SMP6**	I, SG	*Top-It* Card Decks (2 sets of 0–10 Individual Number Cards: *Math Masters*, pp. TA34–TA35)

*Possible Formats/ Groupings: **I** = Individual, **SG** = Small Group, **WC** = Whole Class

Task #			Content Assessed	Possible Formats/ Groupings*	Materials Needed
BOY	MY	EOY			
	10	9	addition number stories K.OA.1, K.OA.2, SMP1, SMP2, SMP4, SMP5	I	counters (at least 10); number line; pencil and paper
	11	10	subtraction number stories K.OA.1, K.OA.2, SMP1, SMP2, SMP4, SMP5	I	counters (at least 10); number line; pencil and paper
	12	11	decomposing numbers K.OA.3, SMP1, SMP2, SMP5	I	**Mid-Year** 20 connecting cubes (10 in each of 2 colors); Building the Number 7 (*Math Masters*, p. 66); crayons (same colors as connecting cubes)
					End-of-Year blank name collection box for the number 9
	13	12	number pairs that add to 10 K.OA.4, SMP1, SMP2, SMP7	I, SG	**Mid-Year** Blank Ten Frame (*Math Masters*, p. TA12); Ten Frame Recording Sheet (*Math Masters*, p. TA19); set of 0–10 Individual Number Cards (*Math Masters*, pp. TA34–TA35); 20 bear counters (10 each of 2 colors); crayons or markers
					End-of-Year 10 bear counters; cup; pencil and paper
		13	addition fluency K.OA.5, SMP6	I, SG	*Dice Addition* dice (dice labeled 0, 0, 1, 1, 2, 2 and 1, 1, 2, 2, 3, 3); Blank Ten Frame (*Math Masters*, p. TA12); markers
		14	subtraction fluency K.OA.5, SMP6	I, SG	*Dice Subtraction* dice (2 dice labeled 0, 1, 2, 3, 4, 5); Blank Ten Frame (*Math Masters*, p. TA12); markers
	14	15	composing teen numbers K.NBT.1, SMP1, SMP2, SMP7	I	**Mid-Year** none
					End-of-Year bag of 15 craft sticks; rubber band; pencil and paper
	15	16	describing and comparing size K.MD.1, K.MD.2, SMP6	I	pair of different-size objects
8			size comparison K.MD.2, SMP6	I	a pair of same-type, but different-size items (e.g. a large and small toy car or ball)

*POSSIBLE FORMATS/ GROUPINGS: **I** = INDIVIDUAL,   **SG** = SMALL GROUP,   **WC** = WHOLE CLASS

Interim Assessments

Task #			Content Assessed	Possible Formats/ Groupings*	Materials Needed
BOY	MY	EOY			
	16	17	sorting **K.MD.3, SMP6, SMP7**	I	attribute blocks
	17		positional language **K.G.1, SMP6**	I, SG	1 bear counter; cup
	18		2-D shape identification **K.G.1, SMP6, SMP7**	I	2-dimensional shapes visible in the environment
		18	identifying 2-D and 3-D shapes, positional language **K.G.1, SMP6, SMP7**	I	2-dimensional and 3-dimensional shapes visible in the environment
9			2-D shape identification **K.G.2, SMP3, SMP6, SMP7**	I	Shape Cards A, B, D, G, I, L, T, U, V (*Math Masters*, pp. TA5–TA10)
	19	19	describing 2-D shapes **K.G.2, K.G.4, SMP3, SMP6, SMP7**	I	Shape Cards B, H, K, M, O, Q, T, U (*Math Masters*, pp. TA5–TA10)
		20	describing 3-D shapes **K.G.2, K.G.4, SMP3, SMP6, SMP7**	I	spherical ball, cubic die, cylindrical die, rectangular prismatic tissue box (or other real-world objects in these shapes)
	20	21	comparing 2-D shapes **K.G.4, SMP6, SMP7**	I	Shape Cards G, I, M, O, T, V (*Math Masters*, pp. TA5–TA10)
		22	comparing 3-D shapes **K.G.4, SMP6, SMP7**	I	objects used for Task 20 (see above)
	21	23	drawing 2-D shapes **K.G.5, SMP4**	I, SG, WC	pencil and paper
		24	building shapes **K.G.5, SMP4**	I, SG	toothpicks; mini-marshmallows
	22	25	composing shapes **K.G.6, SMP1**	I, SG	**Mid-Year** pattern blocks; Pattern-Block Puzzles 1, 2, 3 (*Math Masters*, pp. TA39–TA41); Pattern-Block Puzzles 7, 8 (*Math Masters*, pp. TA45–TA46) (optional)
					End-of-Year set of shapes from *Math Masters*, p. TA55

*Possible Formats/ Groupings: **I** = Individual, **SG** = Small Group, **WC** = Whole Class

NAME _____ DATE _____

Beginning-of-Year Assessment

Administer this face-to-face assessment as early as possible in the school year to gather baseline data to inform your instruction. You can use the extra space below each response to take notes on children's errors and the strategies they use during the task. See pages 8–10 for materials and more information.

Beginning-of-Year Tasks	Responses
1. **Individual** Ask children to count aloud as high as they can, starting at 1. Stop them when their counting becomes erratic. If they stop on their own, ask if they can go higher. Note the highest number children reach before they stop or their counting becomes erratic.	Counts by 1s to _____. *Errors:* **K.CC.1, SMP6**
2. **Individual** Ask children to start at 5 and count until you tell them to stop. Note whether they need to count up to 5 before saying 6, and whether they can count to 10 or higher. If they can count on from 5 to 10, choose a higher starting number and have children count on again.	Counts on from 5 to _____. *Errors:* **K.CC.2, SMP6** Counts on from _____ to _____. *Errors:* **K.CC.2, SMP6**
3. **Individual or Small Group** Give children a plastic bag containing number cards 0–5. Say a number and tell children to reach in, find the number, and hold it up. (You might say "Ready, set, show!" to set a playful tone and ensure that everyone pulls out their cards at the same time.) Repeat until you have called each number in the bag. Add number cards 6–10 for children who are ready. Circle the numbers children recognize and record any errors they make.	Identifies 0 1 2 3 4 5 *Errors:* **K.CC.3** Identifies 6 7 8 9 10 *Errors:* **K.CC.3**
4. **Individual** Gather number cards 0–5 and Quick Look Cards 1, 2, 8, 10, and 12 (from the manipulative kit or connectED.mheducation.com). Make a blank card to represent 0 dots. Display all the number cards so children can see them. Then, one at a time (in random order), show them the dot images. Have children find the number card that matches the dot card. Circle the numbers children match correctly and record any errors they make.	Matches sets of dots to numerals: 0 1 2 3 4 5 *Errors:* **K.CC.3, SMP2, SMP6**

Beginning-of-Year Assessment **11**

NAME   DATE

Beginning-of-Year Assessment (continued)

Beginning-of-Year Tasks	Responses
Individual **5A.** Place 7 connecting cubes in a row. Ask: *How many cubes are there?* Note whether children: • count with number names in the standard order, • pair each cube with only one number name (one-to-one correspondence), and • recognize that the last number they counted tells how many cubes (cardinal principle).	Counts with the correct count sequence: Yes No **K.CC.4a, SMP6** Counts with one-to-one correspondence: Yes No **K.CC.4a, SMP6** Answers "how many?" correctly: Yes No **K.CC.4b, K.CC.5, SMP6**
5B. Spread out the row of 7 cubes as children watch. Ask: *How many cubes now? How do you know?* Note whether children can explain that there are 7 cubes because you did not add or take away any.	Says there are 7 cubes in new arrangement without re-counting: Yes No *Explanation:* **K.CC.4b, SMP6**
5C. Place the 7 cubes in a cup. Say: *How many cubes in the cup? I am putting one more cube in the cup. Now how many cubes are there? How do you know?* Note whether children recognize that there are 8 cubes without re-counting because 8 is one more than 7.	Knows there are 8 cubes without re-counting when one is added: Yes No *Explanation:* **K.CC.4c, SMP6**
6. **Individual** Give children a bag with 20 connecting cubes. Say: *Show me 5 cubes.* Note whether children count out 5 cubes and the strategies they use to keep track of their counting. You may wish to repeat with other numbers of cubes (up to 20) until the task becomes too challenging.	Counts out 5 cubes: Yes No *Errors or strategies:* **K.CC.5, SMP6** Counts out _____ cubes: Yes No *Errors or strategies:* Counts out _____ cubes: Yes No *Errors or strategies:* **K.CC.5, SMP6**

12   Assessment Handbook

NAME   DATE

Beginning-of-Year Assessment (continued)

Beginning-of-Year Tasks	Responses
7. Individual Show children 5 cubes of one color and 2 of a different color. Ask: *Which color has more? How do you know?* Note whether children recognize that 5 is more than 2 and what strategy they use to determine this.	Identifies 5 objects as more than 2 objects: Yes No *Strategy and explanation:* **K.CC.6**
8. Individual Give children a pair of same-type but different-size items (a large and small toy car, for example). Ask: *Which is smaller? Which is bigger?* Note whether they can identify the smaller and bigger object correctly. To make the task more difficult, use objects that are close in size, such as straws or pencils of slightly different lengths.	Identifies object that is *smaller*: Yes No Identifies object that is *bigger*: Yes No **K.MD.2, SMP6**
9. Individual Collect the following Shape Cards from Lesson 1-12: A, B, D, G, I, L, T, U, and V (*Math Masters,* pages TA5–TA10). Spread them out faceup and have children remove shapes as you name them. (Once children have removed the cards for a given shape, have them return all the cards to the table and continue with the next shape.) Say: • *Give me all the circles. How do you know they are circles/this is a circle?* • *Give me all the triangles. How do you know?* • *Give me all the rectangles. How do you know?* • *Give me all the squares. How do you know?* Note which shapes children identify correctly and incorrectly (record letters of the cards they choose; note that the square is a correct choice for the rectangle prompt), and how they identify them (by comparing them to something else, by counting sides or angles, or some other way). Some children may identify prototypical shapes, such as those on Cards A, B, G, T, and V, but not less familiar examples, such as those on Cards I, L, and U.	Circles—correctly identified: _____ Circles—incorrectly identified: _____ *Explanation:* Triangles—correctly identified: _____ Triangles—incorrectly identified: _____ *Explanation:* Rectangles—correctly identified: _____ Rectangles—incorrectly identified: _____ *Explanation:* Squares—correctly identified: _____ Squares—incorrectly identified: _____ *Explanation:* **K.G.2, SMP3, SMP6, SMP7**

Beginning-of-Year Assessment   **13**

NAME _____ DATE _____

Mid-Year Assessment

Administer this face-to-face assessment after Section 5 to assess children's progress on the topics covered through Section 5. Cells shaded in gray are optional because they exceed mid-year expectations for Kindergarten. You can use the space below each response to take notes on children's errors and the strategies they use during the task. See pages 8–10 for materials and more information.

Mid-Year Tasks	Responses	Assess Progress (Circle one.)
Individual **1A.** Ask children to count aloud as high as they can, starting at 1. Stop them when they reach 50 or when their counting becomes erratic. If they stop on their own, ask if they can go higher. Record the highest number they count to successfully. **Look for children to count by 1s through 50.**	Counts by 1s to _____. (50) *Errors:* K.CC.1, SMP6	M N N/A
1B. Prompt children to count by 10s. Stop them when they reach 100 or when their counting becomes erratic. **Look for children to count by 10s through 50.**	Counts by 10s to _____. (50) *Errors:* K.CC.1, SMP6, SMP7	M N N/A
2. Individual Ask children to count on a few numbers from 6, 12, and 35. Note whether they need to count up to the starting number before counting on. Listen especially for difficulties when children count through decade changes. **Look for children to count on from 6, 12, and 35.**	Counts on from 6: Yes No Counts on from 12: Yes No Counts on from 35: Yes No *Errors:* K.CC.2, SMP6	M N N/A
3. Individual Show number cards 0–20 one at a time (in random order) and ask: *What number is this?* Circle each number they can read, and record any errors. You may also wish to write down higher 2-digit numbers to see if children can read them. **Look for children to read numbers 0–20.**	Reads: 0 1 2 3 4 5 6 7 8 9 10 11 12 13 14 15 16 17 18 19 20 K.CC.3 Other numbers read: ___, ___, ___, ___, ___ K.CC.3	M N N/A

Assess Progress   **M** = meeting expectations   **N** = not meeting expectations   **N/A** = not assessed

NAME   DATE

Mid-Year Assessment (continued)

Mid-Year Tasks	Responses	Assess Progress (Circle one.)
4. Individual, Small Group, or Whole Class Collect children's completed number scrolls, or direct children to fill out the first two rows of a blank number scroll. Circle each number they can write, and record any errors. You may wish to save and attach children's scrolls. **Look for children to write numbers 0–20.**	Writes: 0 1 2 3 4 5 6 7 8 9 10 11 12 13 14 15 16 17 18 19 20 Errors: K.CC.3	M N N/A
Individual **5A.** Place 7 connecting cubes in a row. Ask: *How many cubes are there?* **Look for children to:** • count with number names in the standard order, • pair each cube with only one number name (one-to-one correspondence), and • recognize that the last number they counted tells how many cubes (cardinal principle).	Counts with the correct count sequence: Yes No K.CC.4a, SMP6	M N N/A
	Counts with one-to-one correspondence: Yes No K.CC.4a, SMP6	M N N/A
	Answers "how many?" correctly: Yes No K.CC.4b, 4.CC.5, SMP6	M N N/A
5B. Spread out the row of 7 cubes as children watch. Ask: *How many cubes now? How do you know?* **Look for children to explain there are 7 cubes because you did not add or take away any.**	Says there are 7 cubes in new arrangement without re-counting: Yes No Explanation: K.CC.4b, SMP6	M N N/A
5C. Place the 7 cubes in a cup. Say: *How many cubes in the cup? I am putting one more cube in the cup. Now how many cubes are there? How do you know?* **Look for children to recognize that there are 8 cubes without re-counting because 8 is one more than 7.**	Knows there are 8 cubes without re-counting when one is added: Yes No Explanation: K.CC.4c, SMP6	M N N/A

Assess Progress   **M** = meeting expectations   **N** = not meeting expectations   **N/A** = not assessed

| NAME | DATE |  |

Mid-Year Assessment (continued)

Mid-Year Tasks	Responses	Assess Progress (Circle one.)
6. Individual Gather sets of bear counters as described below. Show children the 0–20 number cards. Put out the following sets of bears (being careful not to have children count as you set them out), and ask them to find the card that matches the number of bears. • 5 bears (scattered) • 12 bears (3 rows of 4 bears each) • 20 bears (in a line) • 9 bears (scattered) • 17 bears (in a circle) Circle the number of each set they count correctly, and note their counting strategies. Also circle each numeral they match with the correct set. **Look for children to count correctly and match sets to numerals for all five sets.**	Counts correctly: 5 9 12 17 20 *Strategies:* K.CC.5, SMP6	M N N/A
	Correctly matches sets to numerals: 5 9 12 17 20 K.CC.3, SMP2, SMP6	M N N/A
7. Individual Give children a bag with 20 connecting cubes. Say: *Give me 10 cubes.* Note whether children count out 10 cubes and the strategies they use to keep track of their counting. You may wish to repeat with other numbers of cubes until the task becomes too challenging. **Look for children to count out a set of at least 10 cubes.**	Counts out 10 cubes: Yes No *Errors or strategies:* K.CC.5, SMP6	M N N/A
	Counts out ____ cubes: Yes No *Errors or strategies:* Counts out ____ cubes: Yes No *Errors or strategies:* K.CC.5, SMP6	

Assess Progress   **M** = meeting expectations   **N** = not meeting expectations   **N/A** = not assessed

16 *Assessment Handbook*

NAME   DATE

Mid-Year Assessment (continued)

Mid-Year Tasks	Responses	Assess Progress (Circle one.)
8. **Individual or Small Group** Play several rounds of *Top-It with Dot Cards* (Lesson 2-2) or watch children play together. Note children's ability to compare a variety of sets, including some that are very different (for example, 1 and 7) and some that are closer in number (8 and 9). (Allow children to use counters as supports if desired.) Ask: *Which card has a greater number of dots? Which card has fewer dots? Are these sets equal?* **Look for children to identify sets that are greater, less, and equal.**	Compares sets to determine: Greater: Yes No Fewer (less): Yes No Equal: Yes No *Strategies:* K.CC.6, SMP2, SMP6	M N N/A
9. **Individual or Small Group** Play several rounds of *Top-It with Number Cards* (Lesson 4-12) or watch children play together. Note children's ability to compare various numbers. Ask: *Which number is greater? Which is less? Are these numbers equal?* **Look for children to identify which numeral in a pair is greater and which is less.**	Compares numerals to determine: Greater: Yes No Less: Yes No *Strategies:* K.CC.7, SMP2, SMP6	M N N/A
10. **Individual** Provide children with counters (at least 10), a number line, and pencil and paper, any of which they may use to solve the following parts-and-total number story: *I had 3 green apples and 4 red apples. How many apples did I have all together?* Ask them to represent the story with pictures, fingers, or counters. Note the strategies children use to solve the problem and the ways they represent it. **Look for children to represent and solve the number story concretely with pictures, fingers, or counters.**	Represents number story with pictures, fingers, or counters: Yes No *Representation:* K.OA.1, SMP2, SMP4	M N N/A
	Solves number story: Yes No *Strategy:* K.OA.2, SMP1, SMP5	M N N/A

Assess Progress   **M** = meeting expectations   **N** = not meeting expectations   **N/A** = not assessed

Mid-Year Assessment   **17**

NAME   DATE

Mid-Year Assessment (continued)

Mid-Year Tasks	Responses	Assess Progress (Circle one.)
11. Individual Provide children with counters (at least 10), a number line, and pencil and paper, any of which they may use to solve the following change-to-less number story: *I had 6 stickers and gave 4 to a friend. How many did I have left?* Ask them to represent the story with pictures, fingers, or counters. Note the strategies children use to solve the problem and the ways they represent it. **Look for children to represent and solve the number story concretely with pictures, fingers, or counters.**	Represents number story with pictures, fingers, or counters: Yes No *Representation:* K.OA.1, SMP2, SMP4	M N N/A
	Solves number story: Yes No *Strategy:* K.OA.2, SMP1, SMP5	M N N/A
12. Individual Give children 10 connecting cubes in each of two colors. Have them use both colors to build one tower of 7 cubes. Ask: *How many of each color did you use to make 7? Can you make 7 in another way?* Have children record their combinations (decompositions) on a copy of *Math Masters*, page 66. **Look for children to create at least two different combinations that add to 7 (such as 5 and 2, 4 and 3, or 6 and 1) and to describe and record their combinations.**	Creates at least two combinations for (decompositions of) 7: Yes No *Combinations created:* K.OA.3, SMP1, SMP2	M N N/A
	Records at least two combinations for (decompositions of) 7: Yes No K.OA.3, SMP2	M N N/A
13. Individual or Small Group Play several rounds of *Ten Bears on a Bus* (Lesson 5-3) or watch children play. **Look for children to find number pairs that add to 10.**	Finds number pairs that add to 10: Yes No K.OA.4, SMP1	M N N/A

Assess Progress   **M** = meeting expectations   **N** = not meeting expectations   **N/A** = not assessed

18   Assessment Handbook

NAME   DATE

Mid-Year Assessment (continued)

Mid-Year Tasks	Responses	Assess Progress (Circle one.)
14. Individual Say and write the number 13 and tell children that you will work together to show the number with fingers. Show 10 fingers on your hands and direct them to show additional fingers on their own hands to make 13. Ask them how they knew how many fingers to show. Repeat for the number 17. **Look for children to represent numbers 13 and 17 as 10 and some more fingers.**	Composes and understands 13 as 10 and some more ones: Yes No *Explanation:* Composes and understands 17 as 10 and some more ones: Yes No *Explanation:* K.NBT.1, SMP2, SMP7	M N N/A
Individual **15A.** Give children two objects (one at a time) and say: *Tell me about the length and weight of this object. What else can you tell me about its size?* Record the size language children use. **Look for children to use length and weight language, such as *long, tall, short, heavy,* and *light*, to describe the objects.**	Length and weight language used: Other size language used: K.MD.1, SMP6	M N N/A
15B. After they describe each object individually, ask children to compare the two objects. Ask: *Which is shorter? Which is heavier? How else are these objects similar and different in size?* **Look for children to compare length and weight of objects accurately and to use comparison language, such as *longer, taller, shorter, heavier,* and *lighter*, correctly.**	Identifies *shorter* object: Yes No Identifies *heavier* object: Yes No Other comparison language used: K.MD.2, SMP6	M N N/A

Assess Progress   **M** = meeting expectations   **N** = not meeting expectations   **N/A** = not assessed

NAME _____ DATE _____

Mid-Year Assessment (continued)

Mid-Year Tasks	Responses	Assess Progress (Circle one.)
16. **Individual** Give children a handful of attribute blocks (not a complete set) and ask them to sort the blocks by color. Then have children count the number of blocks in each group and order the groups by count from fewest to most. **Look for children to sort the blocks by color, count the blocks in each group, and order the groups from fewest to most.**	Sorts blocks by color: Yes No Counts blocks in each group: Yes No Orders the groups by count: Yes No K.MD.3, SMP6, SMP7	M N N/A

17. **Individual or Small Group** Give each child a bear counter and a cup. Prompt them to model the following positions: *Place the bear* above *the cup*. Continue to prompt them to place the bear *beside, in front of, next to, below*, and *behind* the cup. You may also provide prompts using other positional words. **Look for children to understand these positional terms and place the bear correctly.**	Positional language understood:
	  above beside in front of
	  next to below behind
	Other positional language understood:
	
	K.G.1, SMP6
		M N N/A

18. **Individual** (Before administering this task, make sure examples of each 2-dimensional shape in the task are visible in the room.) Ask: *Do you see any circles in this room?* Repeat for triangles, squares, and other (non-square) rectangles. Write down the object(s) children identify for each shape. **Look for children to locate at least one example of each of these 2-dimensional shapes in the environment.**	Identifies shapes in the environment: Circle _____ Triangle _____ Square _____ Other rectangles _____ K.G.1, SMP6, SMP7	M N N/A

Assess Progress   **M** = meeting expectations   **N** = not meeting expectations   **N/A** = not assessed

NAME　　　　　　　　　　　　　　　　DATE

Mid-Year Assessment (continued)

19. Individual Collect the following Shape Cards from Lesson 1-12: B, H, K, T, and U (*Math Masters,* pages TA5–TA10). Spread them out faceup and point to one at a time, asking: *What shape is this? How do you know?* Circle the shapes the children identify and name correctly. Record the words children use to explain or describe each shape. **Look for children to identify and name the circle, triangle, square, and rectangle and to use informal language to explain how they identified each one.**	Identifies and names:　　Descriptors for shapes: circle triangle square other rectangle K.G.2, K.G.4, SMP3, SMP6, SMP7	M　N　N/A
You may also wish to show Cards M, O, and Q to see if children can identify a trapezoid, hexagon, and rhombus. Circle the shapes they identify and name correctly and write down the words they use to explain or describe each shape.	Identifies and names:　　Descriptors for shapes: trapezoid hexagon rhombus K.G.2, K.G.4, SMP3, SMP6, SMP7	
20. Individual Show the following pairs of Shape Cards from Lesson 1-12: G and T; T and V; O and M; and G and I (*Math Masters,* pages TA5–TA10). For each pair, ask: *How are these shapes alike? How are they different?* **Look for children to use informal language, as well as mathematical language such as *side* and *vertex*, to describe the similarities, differences, parts, and other attributes of the shapes they compare.**	Language used to describe and compare shapes: *Errors/misconceptions:* K.G.4, SMP6, SMP7	M　N　N/A

Assess Progress　　　**M** = meeting expectations　　**N** = not meeting expectations　　**N/A** = not assessed

NAME _____ DATE _____

Mid-Year Assessment (continued)

Mid-Year Tasks	Responses	Assess Progress (Circle one.)
21. Individual, Small Group, or Whole Class Provide children with pencils and paper. Have them draw a circle, a triangle, a square, and a rectangle. If children seem ready, also ask them to draw a trapezoid, a rhombus, and a hexagon. Circle each shape that they can draw. (You may wish to save and attach children's drawings.) **Look for children to draw recognizable circles, triangles, squares, and rectangles.**	Draws recognizable shapes: circle triangle square rectangle K.G.5, SMP4	M N N/A
	Draws other shapes: trapezoid rhombus hexagon K.G.5, SMP4	
22. Individual or Small Group Provide children with pattern blocks and Pattern-Block Puzzles 1, 2, and 3 (*Math Masters*, pages TA39–TA41). Children should combine shapes to complete these Pattern-Block Puzzles, which contain guiding lines. If they solve the puzzles easily, encourage them to complete the puzzle a different way (by combining shapes to substitute for those outlined on the puzzle) or provide Pattern-Block Puzzles 7 or 8 (*Math Masters*, pages TA45–TA46), which do not contain guiding lines. **Look for children to complete Pattern-Block Puzzles 1, 2, and 3 in at least one way with little or no assistance.**	Completes: Pattern-Block Puzzle 1 Yes No Pattern-Block Puzzle 2 Yes No Pattern-Block Puzzle 3 Yes No Strategies: K.G.6, SMP1	M N N/A
	Completes: Pattern-Block Puzzle 7 Yes No Pattern-Block Puzzle 8 Yes No Strategies: K.G.6, SMP1	

Assess Progress   **M** = meeting expectations   **N** = not meeting expectations   **N/A** = not assessed

NAME _____   DATE _____

End-of-Year Assessment

Administer this assessment near the end of the year to assess children's progress on the topics covered through the end of the program. Double asterisks identify tasks and expectations repeated from the Mid-Year Assessment; administer these tasks only if you need additional information. Cells shaded in gray are optional because they exceed end-of-year expectations for Kindergarten. You can use the space below each response to take notes on children's errors and the strategies they use during the task. See pages 8–10 for materials and more information.

End-of-Year Tasks	Responses	Assess Progress (Circle one.)
Individual **1A.** Ask children to count aloud as high as they can, starting at 1. Stop them when they reach (or just pass) 100 or when their counting becomes erratic. If they stop on their own, ask if they can go higher. Note whether children can count through decade changes (59 to 60, 89 to 90, and so on) and record the highest number they count to successfully. **Look for children to count by 1s through 100.**	Counts by 1s to _____. (100) Errors: K.CC.1, SMP6	M N N/A
1B. Prompt children to count by 10s. Stop them when they reach 100 or when their counting becomes erratic. **Look for children to count by 10s through 100.**	Counts by 10s to _____. (100) Errors: K.CC.1, SMP6, SMP7	M N N/A
2. **Individual** Ask children to count on a few numbers from 12, 35, and 79. Note whether children need to count up to the starting number before counting on. Listen especially for difficulties or errors when children count through decade changes. **Look for children to count on from 12, 35, and 79.**	**Counts on from 12: Yes No **Counts on from 35: Yes No Counts on from 79: Yes No Errors: K.CC.2, SMP6	M N N/A

Assess Progress   **M** = meeting expectations   **N** = not meeting expectations   **N/A** = not assessed

NAME   DATE

End-of-Year Assessment (continued)

End-of-Year Tasks	Responses	Assess Progress (Circle one.)
3. Individual Show number cards 0–20 one at a time (in random order) and ask: *What number is this?* Circle each number they can read, and record any errors. You may also wish to write down other 2- or 3-digit numbers to see if children can read them. **Look for children to read numbers 0–20.	**Reads: 0 1 2 3 4 5 6 7 8 9 10 11 12 13 14 15 16 17 18 19 20 K.CC.3	M N N/A
	Other numbers read: _____, _____, _____, _____, _____, _____, _____, _____, _____, _____ K.CC.3	
4. Individual, Small Group, or Whole Class Collect children's completed number scrolls, or direct children to fill out the first two rows of a blank number scroll. Circle each number they can write, and record any errors. You may wish to save and attach children's scrolls. **Look for children to write numbers 0–20.	**Writes: 0 1 2 3 4 5 6 7 8 9 10 11 12 13 14 15 16 17 18 19 20 Errors: K.CC.3	M N N/A

Assess Progress   **M** = meeting expectations   **N** = not meeting expectations   **N/A** = not assessed

24   Assessment Handbook

NAME   DATE

End-of-Year Assessment (continued)

End-of-Year Tasks	Responses	Assess Progress (Circle one.)
Individual ****5.** Gather sets of bear counters as described below. Show children the 0–20 number cards. Put out the following sets of bears (being careful not to have children count as you set them out), and ask them to find the card that matches the number of bears. • 5 bears (scattered) • 12 bears (3 rows of 4 bears each) • 20 bears (in a line) • 9 bears (scattered) • 17 bears (in a circle) Circle the number of each set they count correctly and note their counting strategies. Also circle each numeral they match with the correct set. **Look for children to count correctly and match sets to numerals for all five sets.**	**Counts correctly: 5 9 12 17 20 *Strategies:* K.CC.5, SMP6 **Correctly matches sets to numerals: 5 9 12 17 20 K.CC.3, SMP2, SMP6	M N N/A M N N/A
6. **Individual** Give children a bag with 30 connecting cubes. Say: *Give me 20 cubes.* Note whether children count out 20 cubes and the strategies they use to keep track of their counting. **Look for children to count out a set of 20 cubes.**	Counts out 20 cubes: Yes No *Errors or strategies:* K.CC.5, SMP6	M N N/A

7. **Individual or Small Group Play several rounds of *Top-It with Dot Cards* (Lesson 2-2) or watch children play together. Note children's ability to compare a variety of sets, including some that are very different (for example, 1 and 7) and some that are closer in number (8 and 9). (Allow children to use counters as supports if desired.) Ask: *Which card has a greater number of dots? Which card has fewer dots? Are these sets equal?* **Look for children to identify sets that are greater, less, and equal.**	**Compares sets to determine. . .
	   Greater: Yes No
	   Fewer (less): Yes No
	   Equal: Yes No
	Strategies:
	
	K.CC.6, SMP2, SMP6
		M N N/A

Assess Progress   **M** = meeting expectations   **N** = not meeting expectations   **N/A** = not assessed

End-of-Year Assessment   **25**

NAME _____ DATE _____

End-of-Year Assessment (continued)

End-of-Year Tasks	Responses	Assess Progress (Circle one.)
8. **Individual or Small Group Play several rounds of *Top-It with Number Cards* (Lesson 4-12) or watch children play together. Note children's ability to compare various numbers. Ask: *Which number is greater? Which number is less? Are these numbers equal?* **Look for children to identify which numeral in a pair is greater and which is less.**	**Compares numerals to determine: Greater: Yes No Less: Yes No *Strategies:* K.CC.7, SMP2, SMP6	M N N/A
9. **Individual** Provide children with counters (at least 10), a number line, and pencil and paper, any of which they may use to solve the following change-to-more number story. *I had 7 stuffed animals and bought 2 more. How many do I have now?* Ask them to write a number sentence to represent the problem. **Look for children to solve the problem correctly and to represent it with a number sentence.**	Represents number story with a number sentence: Yes No *Number sentence:* K.OA.1, SMP2, SMP4	M N N/A
	Solves number story: Yes No *Strategy:* K.OA.2, SMP1, SMP5	M N N/A
10. **Individual** Provide children with counters (at least 10), a number line, and pencil and paper, any of which they may wish to use to solve the following change-to-less number story. *I had 7 grapes and ate 3 of them. How many grapes do I have left?* Ask them to write a number sentence to represent the problem. **Look for children to solve the problem correctly and to represent it with a number sentence.**	Represents number story with a number sentence: Yes No *Number sentence:* K.OA.1, SMP2, SMP4	M N N/A
	Solves number story: Yes No *Strategy:* K.OA.2, SMP1, SMP5	M N N/A

Assess Progress   **M** = meeting expectations   **N** = not meeting expectations   **N/A** = not assessed

NAME _____ DATE _____

End-of-Year Assessment (continued)

End-of-Year Tasks	Responses	Assess Progress (Circle one.)
11. **Individual** Give children a blank name-collection box for the number 9. Prompt them to use drawings or equations (or both) to show at least three different ways to combine numbers to make 9. **Look for children to show at least three different combinations for 9.**	Shows at least three combinations for (decompositions of) 9: Yes No *Names/combinations:* K.OA.3, SMP1, SMP2, SMP5	M N N/A
12. **Individual or Small Group** Play several rounds of *Hiding Bears* (Lesson 6-11) or watch children play together. Have children record the combinations for each round with equations (e.g., 10 = 4 + 6). Note the strategies they use to find the number of hidden bears. **Look for children to find and record the number of bears "hiding" in the cave consistently and accurately.**	Finds number pairs that add to 10: Yes No *Strategies:* K.OA.4, SMP1, SMP7	M N N/A
	Records number pairs that add to 10: Yes No *Number sentences recorded:* K.OA.4, SMP2	M N N/A
13. **Individual or Small Group** Play several rounds of *Dice Addition* (Lesson 7-12) or watch children play together. Record which sums children are able to produce fluently, as well as those that require more time or are incorrect. Also note the strategies children use. If they exclusively use a counting or counting-on strategy, suggest that they try using a quicker strategy or recalling the totals from memory. **Look for children to add small numbers fluently.**	Sums produced fluently: Sums produced slowly or with errors: *Strategies:* K.OA.5, SMP6	M N N/A

Assess Progress   **M** = meeting expectations   **N** = not meeting expectations   **N/A** = not assessed

End-of-Year Assessment (continued)

End-of-Year Tasks	Responses	Assess Progress (Circle one.)
14. **Individual or Small Group** Play several rounds of *Dice Subtraction* (Lesson 8-5) or watch children play together. Record which differences children are able to produce fluently, as well as those which require more time or are incorrect. Also note the strategies children use. If they exclusively use a counting-back strategy, suggest that they try using a quicker strategy or recalling differences from memory. **Look for children to subtract small numbers fluently.**	Differences produced fluently: Differences produced slowly or with errors: *Strategies:* K.OA.5, SMP6	M N N/A
15. **Individual** Give children a bag with 15 craft sticks and tell them how many sticks it contains. First ask them to predict how many bundles of 10 and how many single sticks they will have if they bundle the sticks in groups of 10. Then have them bundle the sticks in this way and write a number sentence to describe their grouping. **Look for children to predict that they can decompose 15 into a group of 10 ones and 5 more ones and then to bundle the craft sticks to show this decomposition. Also look for them to record this grouping with an equation: $15 = 10 + 5$ or $10 + 5 = 15$.**	Predicts that 15 is decomposed as 10 and 5 ones: Yes No Decomposes 15 craft sticks as 10 and 5 ones: Yes No Records composition/decomposition with equation: Yes No *Equation:* K.NBT.1, SMP1, SMP2, SMP7	M N N/A

Assess Progress   **M** = meeting expectations   **N** = not meeting expectations   **N/A** = not assessed

NAME   DATE

End-of-Year Assessment (continued)

End-of-Year Tasks	Responses	Assess Progress (Circle one.)
Individual ****16A.** Give children two objects (one at a time) and say: *Tell me about the length and weight of this object. What else can you tell me about its size?* Record the size language children use. **Look for children to use length and weight language, such as *long*, *tall*, *short*, *heavy*, and *light*, to describe the objects.**	**Length and weight language used: **Other size language used: K.MD.1, SMP6	M N N/A
16B. After they describe each object, ask them to compare the two objects. Ask: *Which is shorter? Which is heavier? How else are these objects similar and different in size?* **Look for children to compare length and weight of objects accurately and to use comparison language, such as *longer*, *taller*, *shorter*, *heavier*, and *lighter*, correctly.	**Identifies shorter object: Yes No **Identifies heavier object: Yes No **Other comparison language used: K.MD.2, SMP6	M N N/A
17. **Individual** Give children a handful of attribute blocks (not a complete set) and ask them to sort the blocks by shape or by size. Then have children count the number of blocks in each group and order the groups by count from fewest to most. **Look for children to sort the blocks by the given attribute, count the blocks in each group, and order the groups by count from fewest to most.**	Sorts blocks by given attribute: Yes No Counts blocks in each group: Yes No Orders the groups by count: Yes No K.MD.3, SMP6, SMP7	M N N/A

Assess Progress   **M** = meeting expectations   **N** = not meeting expectations   **N/A** = not assessed

End-of-Year Assessment   29

NAME _____ DATE _____

End-of-Year Assessment (continued)

End-of-Year Tasks	Responses	Assess Progress (Circle one.)
Individual ****18A.** (Before administering this task, make sure examples of each 2- and 3-dimensional shape in the task are visible in the room.) Ask: *Do you see any circles in this room? Where?* Prompt children to use positional language, such as *above*, *below*, and *next to*, when describing the location of each shape. Repeat for triangles, squares, and other (non-square) rectangles. Write down the object(s) children identify for each shape. Also write down the positional language they use in their descriptions. **Look for children to locate at least one example of each of these 2-dimensional shapes in the environment and use positional language to describe their locations.**	**Identifies 2-D shapes in the environment: Circle _____ Triangle _____ Square _____ Other rectangles _____ K.G.1, SMP6, SMP7	M N N/A
	Positional language used correctly: K.G.1, SMP6	M N N/A
18B. Ask: *Do you see any spheres in this room? Where?* Repeat for cubes, cones, cylinders, and rectangular prisms. Write down the object(s) children identify for each shape. Also write down the positional language they use in their descriptions. **Look for children to locate at least one example of each of these 3-dimensional shapes in the environment and use positional language to describe their locations.**	Identifies 3-D shapes in the environment: Sphere _____ Cube _____ Cone _____ Cylinder _____ Rectangular prism _____ K.G.1, SMP6, SMP7	M N N/A
	Positional language used correctly: K.G.1, SMP6	M N N/A

Assess Progress   **M** = meeting expectations   **N** = not meeting expectations   **N/A** = not assessed

30   Assessment Handbook

NAME   DATE

End-of-Year Assessment (continued)

End-of-Year Tasks	Responses	Assess Progress (Circle one.)
Individual **19A.** Collect the following Shape Cards from Lesson 1-12: B, H, K, M, O, Q, T, and U. Spread them out faceup. Ask: *Are these shapes 2-dimensional or 3-dimensional? How do you know?* **Look for children to identify the shapes as 2-dimensional and to explain that they are 2-dimensional because they are flat.**	Identifies shapes as 2-dimensional: Yes No *Explanation:* K.G.3, SMP3, SMP6, SMP7	M N N/A
19B. Point to one shape at a time (whichever shapes you choose to assess). Ask: *What shape is this? How do you know?* Circle each shape that children can identify and name correctly. Record the words children use to explain or describe each shape. **Look for children to identify and name each shape and to use informal language to explain how they identified it.	Identifies and names: Descriptors for shapes: **circle **triangle **square **other rectangle rhombus trapezoid hexagon K.G.2, K.G.4, SMP3, SMP6, SMP7	M N N/A

Assess Progress   **M** = meeting expectations   **N** = not meeting expectations   **N/A** = not assessed

NAME   DATE

End-of-Year Assessment (continued)

End-of-Year Tasks	Responses	Assess Progress (Circle one.)
20A. Individual Show children a spherical ball, a cubic die, a cylindrical can, and a rectangular-prism tissue box (or other real-world objects in these shapes). Ask: *Are these shapes 2-dimensional or 3-dimensional? How do you know?* **Look for children to identify the shapes as 3-dimensional and to explain that they are 3-dimensional because they are solid and can be held.**	Identifies shapes as 3-dimensional: Yes No *Explanation:* K.G.3, SMP3, SMP6, SMP7	M N N/A
20B. Individual For each object, ask: *What shape is this? How do you know?* Circle each shape at the right that they can identify and name. Record the words children use to explain or describe each shape. **Look for children to identify and name each solid shape and to use informal language to explain how they identified it.**	Identifies and names: Descriptors for shapes: sphere cube cylinder rectangular prism K.G.2, K.G.4, SMP3, SMP6, SMP7	M N N/A
21. Individual Show the following pairs of Shape Cards from Lesson 1-12: G and T; T and V; O and M; and G and I (*Math Masters*, page TA5–TA10). For each pair, ask: *How are these shapes alike? How are they different?* **Look for children to use informal language, as well as mathematical language such as *side* and *vertex*, to describe the similarities, differences, parts, and other attributes of the shapes they compare.	**Language used to describe and compare shapes: *Errors/misconceptions:* K.G.4, SMP6, SMP7	M N N/A

Assess Progress   **M** = meeting expectations   **N** = not meeting expectations   **N/A** = not assessed

32   Assessment Handbook

NAME _____ DATE _____

End-of-Year Assessment (continued)

End-of-Year Tasks	Responses	Assess Progress (Circle one.)
22. **Individual** Show children pairs of real-world objects that are clear examples of geometric solid shapes (ball, die, can, tissue box, cone). For each pair, ask: *How are these shapes similar? How are they different?* **Look for children to use informal language to describe the similarities, differences, parts, and other attributes of the shapes they compare.**	Language used to describe and compare shapes: Errors/misconceptions: K.G.4, SMP6, SMP7	M N N/A

23. **Individual, Small Group, or Whole Class Provide children with pencils and paper. Have them draw a circle, a triangle, a square, and a rectangle. If children seem ready, also ask them to draw a trapezoid, a rhombus, and a hexagon. Circle each shape that they can draw. (You may wish to save and attach children's drawings.) **Look for children to draw recognizable circles, triangles, squares, and rectangles.**	**Draws recognizable shapes:
	  circle triangle
	  square rectangle
	K.G.5, SMP4
		M N N/A

	Draws other shapes: trapezoid rhombus hexagon K.G.5, SMP4	M N N/A
24. **Individual, Small Group, or Whole Class** Provide children with toothpicks and mini marshmallows. Have them build a square and then ask them to build a 3-dimensional shape or structure. **Look for children to build a square and a 3-dimensional shape or structure.**	Builds square: Yes No Builds 3-dimensional shape or structure: Yes No K.G.3, K.G.5, SMP4	M N N/A

Assess Progress   **M** = meeting expectations **N** = not meeting expectations **N/A** = not assessed

NAME   DATE

End-of-Year Assessment (continued)

	End-of-Year Tasks	Responses	Assess Progress (Circle one.)
25.	**Individual or Small Group** Provide a set of shapes from *Math Masters,* page TA55. Prompt children as follows: *Can you show me how to use two triangles to make a rectangle? Can you show me how to use other shapes to make a rectangle? What other shape(s) can you make with these cards?* Note children's strategies for composing shapes. **Look for children to make a rectangle using two triangles and to make a rectangle at least one other way using other shapes.**	Makes a rectangle (or square) from two triangles: Yes No Makes a rectangle from other shapes: Yes No *Strategies:* Other shapes made: K.G.6, SMP1	M N N/A

Assess Progress   **M** = meeting expectations   **N** = not meeting expectations   **N/A** = not assessed

34 Assessment Handbook

NAME _____ DATE _____

Section(s) _____: Individual Profile of Progress

Use this sheet to record children's performance on content or process and practice standards during lesson activities and Assessment Check-Ins.

Lesson	Content and/or Processes and Practices Assessed	Child's Response(s)	Assess Progress	Comments

Assess Progress   **M** = meeting expectations   **N** = not meeting expectations   **N/A** = not assessed

CLASS	DATE

Section(s) ____ : Class Checklist

	Assessment Check-In (or Other) Items by Lesson												
Names													

Assess Progress   **M** = meeting expectations   **N** = not meeting expectations   **N/A** = not assessed

NAME	DATE

Mathematical Process and Practice for Section(s) ___
Individual Profile of Progress

Use this sheet to record children's use of the mathematical processes and practices in lesson activities and Assessment Check-Ins.

	Opportunity	Date	Comments
SMP1: Make sense of problems and persevere in solving them.			
GMP1.1 Make sense of your problem.			
GMP1.2 Reflect on your thinking as you solve your problem.			
GMP1.3 Keep trying when your problem is hard.			
GMP1.4 Check whether your answer makes sense.			
GMP1.5 Solve problems in more than one way.			
GMP1.6 Compare the strategies you and others use.			
SMP2: Reason abstractly and quantitatively.			
GMP2.1 Create mathematical representations using numbers, words, pictures, symbols, gestures, tables, graphs, and concrete objects.			
GMP2.2 Make sense of the representations you and others use.			
GMP2.3 Make connections between representations.			
SMP3: Construct viable arguments and critique the reasoning of others.			
GMP3.1 Make mathematical conjectures and arguments.			
GMP3.2 Make sense of others' mathematical thinking.			
SMP4: Model with mathematics.			
GMP4.1 Model real-world situations using graphs, drawings, tables, symbols, numbers, diagrams, and other representations.			
GMP4.2 Use mathematical models to solve problems and answer questions.			

Record-Keeping Masters

NAME _____ DATE _____

Mathematical Process and Practice for Section(s) _____
Individual Profile of Progress (continued)

	Opportunity	Date	Comments
SMP5: Use appropriate tools strategically.			
GMP5.1 Choose appropriate tools.			
GMP5.2 Use tools effectively and make sense of your results.			
SMP6: Attend to precision.			
GMP6.1 Explain your mathematical thinking clearly and precisely.			
GMP6.2 Use an appropriate level of precision for your problem.			
GMP6.3 Use clear labels, units, and mathematical language.			
GMP6.4 Think about accuracy and efficiency when you count, measure, and calculate.			
SMP7: Look for and make use of structure.			
GMP7.1 Look for mathematical structures such as categories, patterns, and properties.			
GMP7.2 Use structures to solve problems and answer questions.			
SMP8: Look for and express regularity in repeated reasoning.			
GMP8.1 Create and justify rules, shortcuts, and generalizations.			

38 Assessment Handbook

CLASS _____ DATE _____

Mathematical Process and Practice Opportunities
Class Record

Standard for Mathematical Process and Practice _____

Goal for Mathematical Process and Practice _____

Opportunity: _____

Names	+ / ✓ / −	Comments

Record-Keeping Masters

NAME DATE

Mathematical Process and Practice in Open Response Problems
Individual Profile of Progress

Lesson	SMP	GMP	Assess Progress	Comments
2-7	8	8.1		
3-7	2	2.1		
4-7	1	1.5		
5-7	3	3.1		
6-7	5	5.2		
7-7	6	6.3		
8-7	7	7.1		
9-7	4	4.1		

Assess Progress **E** = exceeding expectations **M** = meeting expectations
  P = partially meeting expectations **N** = not meeting expectations

40 *Assessment Handbook*

NAME	DATE

My Exit Slip

- -

NAME	DATE

My Exit Slip

NAME   DATE

Good Work!

☺ I have chosen this work because

_____

_____

_____

_____

_____

_____

_____

NAME _____ DATE _____

My Work

This work shows that I can _____

_____

_____

I am still learning to _____

_____

_____

_____

--- ✂ - ✂ ---

NAME _____ DATE _____

My Work

This work shows that I can _____

_____

_____

I am still learning to _____

_____

_____

_____

| NAME | DATE |  |

About Math Time

Draw a face or write the words that show how you feel.

 Good    OK    Not so good

① This is how I feel about math:	② This is how I feel about working with a partner or in a group:	③ This is how I feel about working by myself:
④ This is how I feel about solving number stories:	⑤ This is how I feel about doing Home Links with my family:	⑥ This is how I feel about finding new ways to solve problems:

Circle **yes**, **sometimes**, or **no**.

⑦ I like to figure things out. I am curious.

   yes   sometimes   no

⑧ I keep trying even when I don't understand something right away.

   yes   sometimes   no

44 Assessment Handbook

GMP1.1 Rubric

SMP1: Make sense of problems and persevere in solving them.

Goal for Mathematical Practice	Not Meeting Expectations	Partially Meeting Expectations	Meeting Expectations	Exceeding Expectations
GMP1.1 Make sense of your problem.				

GMP1.2 Rubric

SMP1: Make sense of problems and persevere in solving them.

Goal for Mathematical Practice	Not Meeting Expectations	Partially Meeting Expectations	Meeting Expectations	Exceeding Expectations
GMP1.2 Reflect on your thinking as you solve your problem.				

GMP1.3 Rubric

SMP1: Make sense of problems and persevere in solving them.

Goal for Mathematical Practice	Not Meeting Expectations	Partially Meeting Expectations	Meeting Expectations	Exceeding Expectations
GMP1.3 Keep trying when your problem is hard.				

Generic GMP Rubrics

GMP1.4 Rubric

SMP1: Make sense of problems and persevere in solving them.

Goal for Mathematical Practice	Not Meeting Expectations	Partially Meeting Expectations	Meeting Expectations	Exceeding Expectations
GMP1.4 Check whether your answer makes sense.				

GMP1.5 Rubric

SMP1: Make sense of problems and persevere in solving them.

Goal for Mathematical Practice	Not Meeting Expectations	Partially Meeting Expectations	Meeting Expectations	Exceeding Expectations
GMP1.5 Solve problems in more than one way.				

GMP1.6 Rubric

SMP1: Make sense of problems and persevere in solving them.

Goal for Mathematical Practice	Not Meeting Expectations	Partially Meeting Expectations	Meeting Expectations	Exceeding Expectations
GMP1.6 Compare the strategies you and others use.				

GMP2.1 Rubric

SMP2: Reason abstractly and quantitatively.

Goal for Mathematical Practice	Not Meeting Expectations	Partially Meeting Expectations	Meeting Expectations	Exceeding Expectations
GMP2.1 Create mathematical representations using numbers, words, pictures, symbols, gestures, tables, graphs, and concrete objects.				

GMP2.2 Rubric

SMP2: Reason abstractly and quantitatively.

Goal for Mathematical Practice	Not Meeting Expectations	Partially Meeting Expectations	Meeting Expectations	Exceeding Expectations
GMP2.2 Make sense of the representations you and others use.				

GMP2.3 Rubric

SMP2: Reason abstractly and quantitatively.

Goal for Mathematical Practice	Not Meeting Expectations	Partially Meeting Expectations	Meeting Expectations	Exceeding Expectations
GMP2.3 Make connections between representations.				

GMP3.1 Rubric

SMP3: Construct viable arguments and critique the reasoning of others.

Goal for Mathematical Practice	Not Meeting Expectations	Partially Meeting Expectations	Meeting Expectations	Exceeding Expectations
GMP3.1 Make mathematical conjectures and arguments.				

GMP3.2 Rubric

SMP3: Construct viable arguments and critique the reasoning of others.

Goal for Mathematical Practice	Not Meeting Expectations	Partially Meeting Expectations	Meeting Expectations	Exceeding Expectations
GMP3.2 Make sense of others' mathematical thinking.				

GMP4.1 Rubric

SMP4: Model with mathematics.

Goal for Mathematical Practice	Not Meeting Expectations	Partially Meeting Expectations	Meeting Expectations	Exceeding Expectations
GMP4.1 Model real-world situations using graphs, drawings, tables, symbols, numbers, diagrams, and other representations.				

GMP4.2 Rubric

SMP4: Model with mathematics.

Goal for Mathematical Practice	Not Meeting Expectations	Partially Meeting Expectations	Meeting Expectations	Exceeding Expectations
GMP4.2 Use mathematical models to solve problems and answer questions.				

GMP5.1 Rubric

SMP5: Use appropriate tools strategically.

Goal for Mathematical Practice	Not Meeting Expectations	Partially Meeting Expectations	Meeting Expectations	Exceeding Expectations
GMP5.1 Choose appropriate tools.				

GMP5.2 Rubric

SMP5: Use appropriate tools strategically.

Goal for Mathematical Practice	Not Meeting Expectations	Partially Meeting Expectations	Meeting Expectations	Exceeding Expectations
GMP5.2 Use tools effectively and make sense of your results.				

GMP6.1 Rubric

SMP6: Attend to precision.

Goal for Mathematical Practice	Not Meeting Expectations	Partially Meeting Expectations	Meeting Expectations	Exceeding Expectations
GMP6.1 Explain your mathematical thinking clearly and precisely.				

GMP6.2 Rubric

SMP6: Attend to precision.

Goal for Mathematical Practice	Not Meeting Expectations	Partially Meeting Expectations	Meeting Expectations	Exceeding Expectations
GMP6.2 Use an appropriate level of precision for your problem.				

GMP6.3 Rubric

SMP6: Attend to precision.

Goal for Mathematical Practice	Not Meeting Expectations	Partially Meeting Expectations	Meeting Expectations	Exceeding Expectations
GMP6.3 Use clear labels, units, and mathematical language.				

GMP6.4 Rubric

SMP6: Attend to precision.

Goal for Mathematical Practice	Not Meeting Expectations	Partially Meeting Expectations	Meeting Expectations	Exceeding Expectations
GMP6.4 Think about accuracy and efficiency when you count, measure, and calculate.				

GMP7.1 Rubric

SMP7: Look for and make use of structure.

Goal for Mathematical Practice	Not Meeting Expectations	Partially Meeting Expectations	Meeting Expectations	Exceeding Expectations
GMP7.1 Look for mathematical structures such as categories, patterns, and properties.				

GMP7.2 Rubric

SMP7: Look for and make use of structure.

Goal for Mathematical Practice	Not Meeting Expectations	Partially Meeting Expectations	Meeting Expectations	Exceeding Expectations
GMP7.2 Use structures to solve problems and answer questions.				

GMP8.1 Rubric

SMP8: Look for and express regularity in repeated reasoning.

Goal for Mathematical Practice	Not Meeting Expectations	Partially Meeting Expectations	Meeting Expectations	Exceeding Expectations
GMP8.1 Create and justify rules, shortcuts, and generalizations.				